AF483838

To my nieces and nephews,
along with honorary nieces and nephews
all over the world.
You are dearly loved!

My Aunt likes plants,
inside and out.
She likes watching things grow,
there is no doubt.

From spider plants to snake plants
-little and big-
some are very bushy,
others look like a twig.

Plants grow all
around the world, you see.
But growing them in a house
takes extra T.L.C.

Some plants like the shade
and some like the sun.
I'm starting to think growing plants
is quite fun!

A cactus has spikes
that hurt if you touch.
Succulent leaves are puffy,
so you can't water much.

Many plants clean
the air that we breathe.
They're not only plants –
they're like employees!

Researchers are testing
if plants can see and hear,
because plants move their
roots to where water is near.

Some flowers are colorful
with petals so bright.
A meadow of wildflowers
makes a beautiful sight!

It's fun to learn
what each plant will need
to grow leaves,
then flowers,
and then to make seeds.

We water the garden
to help food to grow.
While we're there
I chomp
a little tomato.

My Aunt loves to dig
and she's not afraid of bugs.
She really likes bees
but does not like slugs.

Your zone will say
which plants
can grow close outside.
Some grow
just one place in the world,
others grow far and wide.

When you plant your own garden,
you can grow vegetables
you've never seen!
Like a yellow carrot
or a purple green bean.

To be honest,
the only difference
between a garden and a weed
is often how much care
from people it will need.

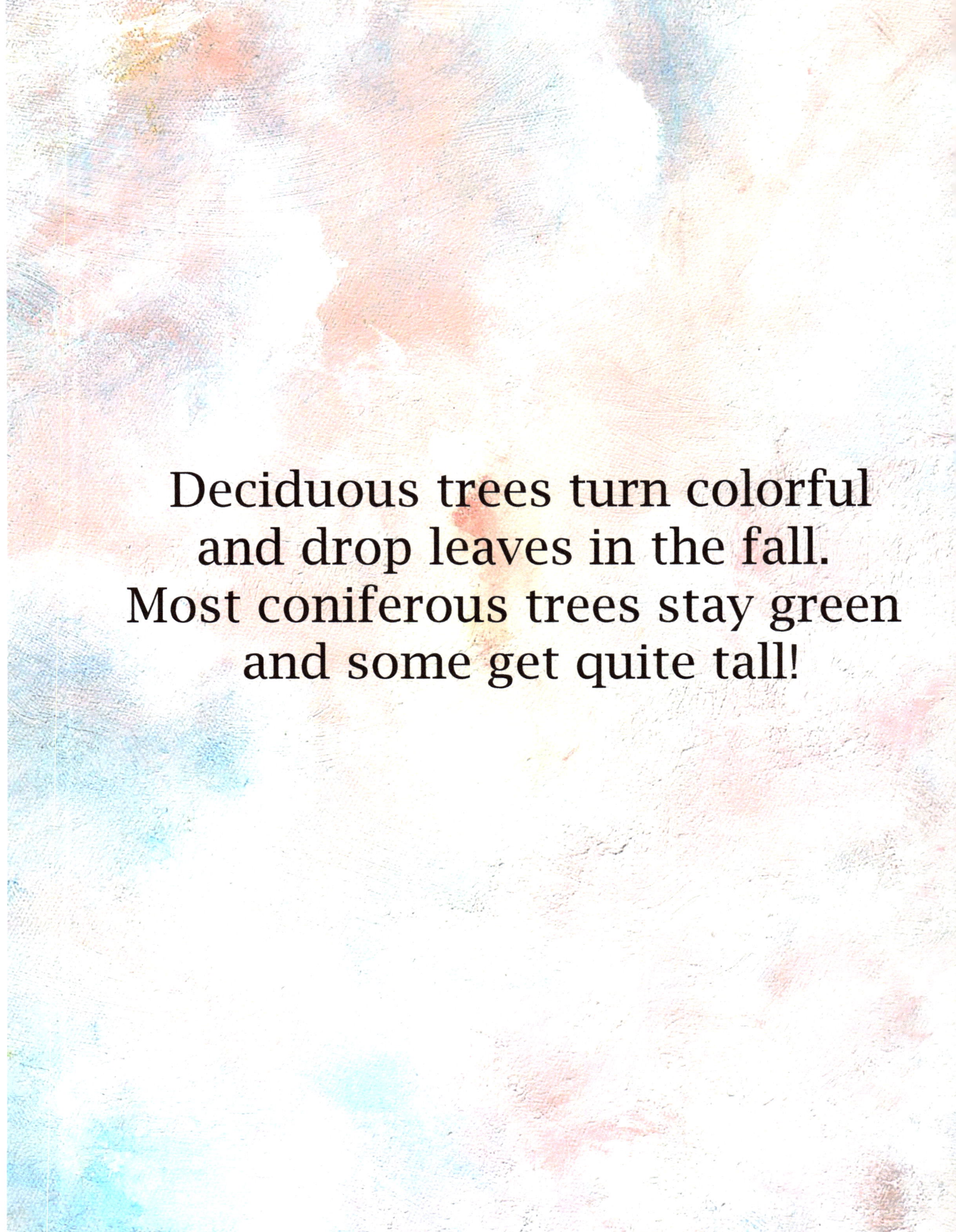
Deciduous trees turn colorful
and drop leaves in the fall.
Most coniferous trees stay green
and some get quite tall!

Plants give animals
the food that they need,
like acorns, nectar
and even bird seed.

I walk dogs with my Aunt
on paths lined with trees,
and we identify
all the animals and plants
that we see.

Not all plants are good
but some can be used
medicinally
like jewelweed
which helps
prevent poison ivy.

Plantain weed can shoo
mosquitoes,
and yarrow salve feels good
on my sore toes.

Wood sorrel tastes lemony
and is quite a delight.
But don't eat all plants–
some plants are not nice!

Anyone can grow plants,
you don't need special powers!
You can grow them for food,
just for fun,
or for flowers.

My Aunt likes
watching things grow, you see,
but what she loves
watching grow most
is ME!